Nicholas G. Stangarone

The Ring of Power

Gold Mind Press

The Ring of Power

Copyright © 2014 - 2017 by Nicholas G. Stangarone

All Rights Reserved. No part of this book may be reproduced or transmitted in any form or by any means, electronic or mechanical, including photocopying, recording, or by any information storage and retrieval system without permission in writing from the publisher.

ISBN-13: 978-1502976338

ISBN-10: 1502976331

Printed in the United States of America

Second Printing 2017

Dedicated to my Father.

For he inspired the spark
        that cast the light from which I saw this,

    "Ring of Power."

Contents

Introduction *page 6*

CHAPTER ONE
Reaping the Whirlwind. *page 8*

CHAPTER TWO
Thunderbolts in a Whirlwind. *page 16*

CHAPTER THREE
What is the Ring of Power? *page 19*

CHAPTER FOUR
New Ordered World is NOW. *page 32*

Bibliography *page 38*

Introduction

To reap the whirlwind and harness its potential for the benefit of Earth's inhabitant's is at the heart of what makes the "Ring of Power," truly worth realizing. And this is the promise that is presented to anyone willing to consider the merits of this new energy production stratagem.

But before delving into the operational characteristics of the "Ring of Power," the first two chapters are to form the groundwork for how and why this invention can offer humanity the promise of a most virtuous future.

The intentions for writing this book are not to provide any additional research on the physics of angular momentum, a subject already saturated with reams of scholarly research. Nor is this book intended to offer any additional insights or specifications toward the design of alternator to be utilized for this invention.

However, by offering an accurate account of the "Ring of Power" and its performance capabilities, one may be afforded a richer appreciation for the awesome ramifications of addressing society's modern incongruities on one united front.

Specifically, the "Ring of Power," promises to offer the following benefits:

1.) Renewably abundant electrical power.

2.) Cleaner and healthier environment.

3.) Physically fitter and healthier population.

4.) Zero unemployment.

5.) Enhanced camaraderie and communal spirit.

Momentum is a physical property that permeates all forms of "mass in motion" relative to a constant gravitational force field. This fundamental property of motion can be transferred or shared between various states of mass. Momentum can also be leveraged to multiply the effects of this fundamental force. And therefore "leveraged momentum" has the potential to provide a new found freedom in electrical power generation. Still, it's true potential has, as of this latest printing, yet to be tested.

No later than 2019 or ten years from the conception of the original "Ring of Power," the preliminary testing of this revolutionary invention is planned to proceed in a yet undisclosed location. While the design will be kept secret at this point in time, its codename is, *"Cyclone Caesar."*

Chapter One

Reaping the Whirlwind

The two underling principles powering this electrical generating invention are leverage and momentum. Out of these two, the former was first to be adequately described in purely mathematical terms. And one could argue that Archimedes, the great mathematician and inventor from antiquity etches a clearly defined line of demarcation where thereafter the mechanical advantage of the lever was fully understood for all its useful effectiveness.

The reason for this is simple and evidenced by the fact that Archimedes is most remembered for his claim that if given a suitably sized fulcrum, he could move Earth itself. Where to would be difficult to speculate, but it can assumed that perhaps most of the laws and formulas dealing with leverage were understood and established by the year 212 BC, the year of Archimedes' assassination.

But more importantly, the incomparable Archimedes is credited the discovery of the so called, "law of the lever." And the mathematical equation that describes this mechanical advantage is given by the fact that the output of physical effort exerted is divided by the input of effort or force applied is equal to the proportional dimension of the leveraged input divided by the leveraged output:

$$Fo / Fi = Li / Lo.$$

Why this equation is so important is because it demonstrates in undeniable mathematical terms that leverage, in and of itself is equivalent to force. Now the reason why leveraged was understood over a thousand years before momentum, at least mathematically speaking, appears to indicate that leverage, a static phenomena, conforms rather ideally to the defining rigors of geometry.

On the other hand, momentum is essentially a dynamic quantity relative to the speed or velocity of the moving object's specific quantity of mass. This kinetic state is too dynamic to present itself to any tactile grasp; it's fleeting consistency is unvariably lost by the "forced" measurement. This stubborn inability to hold momentum in the palm of one's hand simply hampered Aristotle's philosophic appreciation of the motion's dynamic characteristics.

And since Aristotle's scientific principles were deemed above reproach for hundreds of years after the Greek philosopher's death, it may have also acted to forestall any substantial attempts to discover how to adequately measure momentum.

Anyway, history records that it wasn't until the fall of the Western Roman Empire that men of courage and intellect dared to present logical arguments against the prevailing misconceptions.

As such, Euclid's spacial geometry and Aristotle's insightful physics would not yet meld with Archimedes' intuitive understanding of leverage and thereby establish the requisite intellectual underpinnings that proved necessary for a fairly reasonable understanding of motion.

The result was that numerous centuries would come

and go before the latent scientist or sixth century philosopher, John Philoponus offered insights so profound that they eventually gained acceptance as impetus, inertia and finally momentum.

Since the transparent dynamism of motion proved so elusive to grasp intellectually, it subsequently challenged ingenious minds to devise ever clever means of calculating momentum's constantly changing magnitude.

One of these deceptively simple methods involved dropping heavy objects from various heights directly onto prepared lumps of clay. The greater the depression or depth of crater sustained by each impact presented a proportional impression of the collision's force. By measuring and analyzing the differences of each crater's depth, they could begin to develop a primitive, though highly tactile understanding of this dynamic phenomena, if only by proxy.

By the 14th century, Jean Buridan had deciphered a fundamental characteristic about "impetus" that set in motion the intellectual drive to finally capture this kinetic phenomena. J.B.'s brilliant counterpoint to the whole Aristotelian motion theory was in stating with unequivocal fervor that the force "imparted" to overcome an object's inertial mass and thus set the object into a "new motion" does "not diminish" spontaneously.

This stunning revelation was further defined by other Buridan disciples or contemporaries like Nicole Oresme and Simon Stevin. In time, each of these two men were to develop highly intuitive methods of analysis that effectively cleared the murky haze of antiquated thought and conjecture from the discerning logic of scientific reasoning.

It's been over 450 years since Galileo Galilei was born on the Italian peninsula. When the notion of momentum was constrained by the static measurements of motion. But this man's incredible genius would conjure such exquisite experiments of motion that they allowed him to ascertain a tactile understanding of this universal dynamism which heretofore had been well beyond the scope of any other's previous comprehension.

Yet with all his ingenuity, Galileo still lacked the mathematical tools and precision time keeping instruments to accurately measure the constantly changing values. Therefore, Galileo was never fully able to interpret the scientific data from real world experiments he'd carried out with the diligence of a true scientist while still a young man. However, his efforts would eventually inspire others to formulate several universal laws of physics.

Unfortunately for our tragic hero, Galileo, his brilliant trajectory would be cruelly impeded by the severely threatened theologian dogma of that era.

And so it would fall onto the curious and calculating minds of others, most especially, one who rather fatefully had been born the same year of Galileo's death. When the arc of discovery passed suddenly from Italy to England and eventually to Trinity College Cambridge and its mathematical theologian, Sir Isaac Newton.

Equipped with the computational agility of his calculus equations, Newton succeeded in subduing the elusive genie by subscribing dynamic quantity to the kinetic transparency of momentum.

This calculating prowess was soon blustered by Euler,

when he added his mathematical genius to this age-old quest. Thereafter, the kinetic quantities of motion could be examined within the full grasp of man's imagination.

And it was from this formidable convergence of intellect that the lion's share of those mathematical formulas necessary for calculating that universal dynamism known as momentum were derived. And as such, momentum could now be stated mathematically as the product of an object's mass (gravitational constant) multiplied by the speed (delta x) at which the object's mass travels:

$$\textbf{Momentum = Mass x Velocity}$$

$$\textbf{P = M x V}$$

To gain a fuller understanding, let's consider this excerpt from an article titled, "Football Physics: The Anatomy of a Hit." The article was downloaded from the "Popular Mechanics" website. Basically it addressed the force equivalent of momentum in a highly imaginative manner, "At 5 ft. 11 in. and 199 pounds, Marcus Trufant is an average-size NFL defensive back (DB) ... in a league where more than 500 players weighed 300-plus pounds at the 2006 training camps. But a DB's mass combined with his speed -- on average, 4.56 seconds for the 40-yard dash -- can produce up to 1600 pounds of tackling force, according to Timothy Gay, a physics professor at the University of Nebraska and author of The Physics of Football."

So, even though the football player's weight never changed throughout the entire duration or course of the forty yard dash, his traveling weight was able to generate a force equaling 1600 pounds! Still, when he eventually slowed down to a stand still, all that force was lost or completely dissipated into thin air.

Wasn't this not the same explosive power carried by a charging calvary and delivered upon an opposing army's infantry flanks to disrupt their offensive formations? What about those onslaughts of warriors, as they charged pell-mell from higher ground on the standing enemy combatants down below? In either circumstance, the end result is calculable because of the conservation of momentum.

The following insight comes to us from the website at, "www.faqs.org/sports-science/Fo-Ha/Football-Mass-Momentum-and-Collisions.html," hopefully it provides additional mathematical insight to gauge or ascertain the possibilities.

"A 240 pound linebacker hits a 240 pound fullback who is running at 9 yards per second. After the collision, the fullback's final speed is zero.

From the time the linebacker first touches the fullback to the time his forward motion is stopped is about 0.2 second. The deceleration is -135 ft per second^2.

When the fullback's mass (240 lbm) is multiplied by the deceleration (-135 ft/sec^2) the resulting net force is no less than 32,400 lbm-ft/sec squared. A force is equivalent to about one-half ton (½ ton) in the negative (backward) direction.

Isaac Newton's third law of motion states that for every action, there is an equal and opposite reaction. By virtue of their weight, both football players will exert the same force on each other during the collision and continue to do so over the same time interval."

But the important difference here is that its in opposite

directions. Because of this one player gains the same momentum (mass × velocity) that the other player loses. The net change in the momentum of the two players is zero. This is called conservation of momentum.

However, the force described here exists on a linear plane. Whereas the forces most prevalent with the "Ring of Power," act on a rotating circular plane. But before we touch on angular momentum, we'll consider the action that interacts with both planes of force. This is what is known as torque or more specifically, "torque about an axis" and it is predominantly a linear action that induces angular motion.

In physics, the Greek letter "tau," is used to signify torque and is basically the product of the "force applied" multiplied by the "length of the lever arm" (D) to its axis. Remember the "law of the lever?" Essentially this is the force that turns the "Ring of Power."

Torque = F x D
Torque = Moment of Inertia (x) Angular Acceleration.
Angular Momentum = Moment of Inertia (x) Angular Velocity.

Once the "Ring of Power," is set to motion, the angular dynamics immediately come into play. Specifically though, it is angular momentum that concerns us most because this force maintains the motion and torque required to continuously rotate the "Ring of Power."

So with a minimal amount of input through pedaling by the bicyclists, their speed can be sustained for long periods of electrical production.

The equation for angular momentum shows that it's a product of an object's inertial mass and only it's attained velocity and not its acceleration! But more importantly, it mathematically punctuates the fundamental difference in using a stationary bike versus an ordinary bicycle to generate electricity. Specifically, there is the force required to attain a certain speed or acceleration and then there is the force necessary to maintain the identical distance of space per duration of time or velocity.

To appreciate what that implies, one need only consider what happens when trying to use a stationary bike as a prime mover to generate electricity. The cyclist not only quickly reaches a state of exhaustion trying to maintain the necessary rpms, but eventually they will overheat without the adequate ventilation. But more importantly, the moment they stop to pedal their brains out, all power is lost.

Whereas when you stop pedaling on a moving bicycle and begin to coast, the momentum you had generated with your leg muscles is converted to inertia that will carry you forward at the same speed till it is gradually dissipated by wind resistance, gravity and friction. The moment you stop pedaling does not result in an immediate loss of power.

Also without the conservation of angular momentum, Olympic figure skaters could never perform those exhilarating spins that turn to gold before the eyes of the world. That's because when they begin their spin with arms extended, they capture a velocity of rotation pertaining to a specific moment of inertia. But the conservation of angular momentum "conserves" the force or momentum, there the rate of rotation must be accelerated if the skater retracts their arms into a tighter circle or radius. So, by reducing their body's moment of inertia about an axis and angular momentum remains constant, then the speed of rotation can only increase!

Of course, there are other properties of angular momentum that contribute to this phenomena, but it is beyond the scope of this book. It is only important that you recognize the significance of this speed-sustaining phenomena, which has hitherto been underappreciated as well as underutilized.

Chapter Two

Thunderbolts in a Whirlwind

The electric spark has always fascinated our natural curiosity. But the cognitive ability to ascertain the source of this phenomena was until recently only rudimentary at best. And as the ancient mind harbored a proclivity or propensity to ascribe the unexplainable to the celestial realm of the gods, the awesome power of the thunderbolt was thus appropriated to the arsenal of those worthy immortals.

While there exist evidence that the electric spark had been contained by man's ingenuity as early as the time of the invention of the "Baghdad" battery. Whether these archeological artifacts were ever used for the purpose of producing an electrical potential or voltage is of negligible concern.

This is because what's truly important here is substantiated evidence or lack thereof that the western world would not come close to developing a similar apparatus for inducing an electrical current until the time of the Enlightenment. When men such as Galvani, Volta and the statesman/ inventor Franklin would each usher the study of electricity beyond a parlor trick to thereafter enter the lofty realm of scientific inquiry.

While there exists evidence that the interaction of electricity and magnetism may have been recognized as early as 1802 by Gian Domenico Romagnosi, it wouldn't be until 1820 when Hans Christian Orsted brought the discovery of the eternal relationship between electricity and magnetism to the forefront of western intellectual inquiry.

When this insight sparked Michael Faraday to build new and fascinating electrical devices, the reams of empirical data wrought from the testing of these prototypes eventually provided James Clerk Maxwell all he needed to formulate the fundamental laws of electromagnetism.

And because of Maxwell's masterful mathematical formulas, people could effectively grasped the spark between their fingertips of their minds. Pandora's box had been open and electrified, thus inviting truly brilliant minds to reach in and pull out amazing new technologies. One of these personages was the electrical wizard, Nikola Tesla.

And while practical electrical generators had been previously introduced by a number of inventors, namely J.E.H. Gordon, Lord Kelvin and Sebastian Ferranti, Tesla's designs and inventions for the polyphase alternator are what moved us into the modern world. Because by the time Nikola Tesla had achieved his childhood dream of using the awesome flowing power of Niagara Falls to generate electricity, he had provided the fundamental schematics for most of the alternators used today, well over a full century later.

While the automobile alternator is probably the most recognizable electrical generator, there are hundreds of other design variations to handle any number of applications from the locomotive diesel engine to the ubiquitous wind turbine.

Today, alternators are classified not only according to

its applications, but also by the alternator's mode of excitation, rotation and phase. However this will not impede our intentions to develop and design new alternators for the specific use of the "Ring of Power."

Aside from that, any additional information concerning the alternator is beyond the scope of this edition. However, further editions may elaborate further on the different alternators and their peculiar attributes for suitability to the operation of the "Cyclone Caesar."

Chapter Three

What is the Ring of Power?

On November 1st of the year 2009, while contemplating various rotating bases to improve on the design of a sail driven windmill which I'd first imagined around the year 1980 and a quarter of a century later crafted a working prototype of the unorthodox windmill design. Yet, however lacking the prototype may have appeared in construction, it did demonstrate some intriguing and also promising properties of sustained rotation.

But before I could do any follow-up improvements, life threw me a curve ball, causing me to move cross country, not once but twice in five years!

As it was, on my third return to California, which was followed by a few years of desperate financial hardship brought on by the Great Recession, I attempted to redesign a new working prototype of the sail 'n windmill which would incorporate some of the improvements I had learned from conducting the 2005 prototype test. Undoubtedly, this renewed effort was spurred on by the record prices for gasoline and the demand for alternative energy sources.

As I began to imagine different versions of a revolving base that would address the stress and power requirements, a unique design with intriguing fractal properties came to mind. After developing a more robust version of the initial thought in my mind, I decided to abandon the notion of developing the sail'n windmill prototype, because suddenly it had become glaringly apparent that from this point on, the new electrical generating invention could be driven by other means. That is, besides wind as motive source of propulsion, now leveraged momentum could be used to sustain speed.

Why bicycles? Well, first off, I've been an avid bicyclist for nearly half a century. But perhaps more significantly, at the start of this new millennium, I assembled a stationary bike electrical generator to test it's potential. But upon the first tests of this contraption, it became immediately apparent that something was missing.

Most probably, it was because the power of my pedaling that I'd experienced while bike riding, could not be realized while pedaling my heart out on the stationary bike. Unable to realize at the time how to reconcile this deficiency, I put any effort to generate electricity by pedal power to the wayside.

So, over the next years, it remained on the back burner as I gradually worked my way out of my financial nightmare. But occasionally, when time permitted, I would do research on the design, making improvements along the way as I gained further knowledge about this new and original approach to generating electrical power.

Over a year ago I completed a provisional patent application for the original model. Still circumstances would not allow to proceed as planned and it was put on the back burner once more.

But in honor of the late great Steve Jobs, I purposely waited until the month of October, or more specifically, Wednesday, October 1, 2014, to submit a United States

Provisional Patent application for the "Ring of Power." So I can therefore state that by October 5th, the invention was patent pending.

The important aspects of the provisional patent application are as follows:

Brief Summary of the Invention

The **Cyclonic Cylinder STS (Speed Transfer System)** is a leveraged momentum machine that transfers or converts the speed of the traveling cyclists into the sufficient rpm of the alternator rotor to produce electrical power.

Basically, a minimum of four to six PowerWheels are position at equal distance from a center point and also at equal spacing from each other. Each PowerWheel serves three purposes. First they are to act as supports for the integrity of the circular form. Secondly, there are to secure at least one alternator to the post and the power wheel pulley. Thirdly, they are to transfer the rotational speed of the Transfer Cylinder to the alternator rotor.

Once the power wheels are secured and the transfer cylinder assembled, the **Cyclonic Cylinder STS (Speed Transfer System)** can be operated by a minimum of two bicyclists riding in tandem and positioned at opposite ends of the Transfer Cylinder.

The momentum of the bicyclists is leveraged on the transfer cylinder by the rapid displacement of their polar positioning (180 degree separation) on the transfer cylinder. Also the flywheel effect of the Transfer Cylinder coupled with the angular momentum of the bicyclists will act to maintain the acquired speed.

Riding in tandem to maintain a speed between 15-30 mph, revolves the Transfer Cylinder at the optimal speed to spin the minimum number of four alternators at the rpms necessary to generate electricity of sufficient voltage and current to power the basic needs of its users.

In effect, by riding in tandem, a constant speed can be maintained for extended periods and more efficiently than riding individually.

Brief Description of the Drawings

Drawing No.1: *Diagram of the Cyclonic Cylinder STS (Speed Transfer System).*

Figure 1 shows diagram of the **Cyclonic Cylinder STS (Speed Transfer System)** in the basic configuration. This basic configuration requires a minimum number of PowerWheels (4) and bicyclists (2).
1.) **Transfer Cylinder**
2.) **PowerWheel with attached Alternator**
3.) **Standard Bicycle**

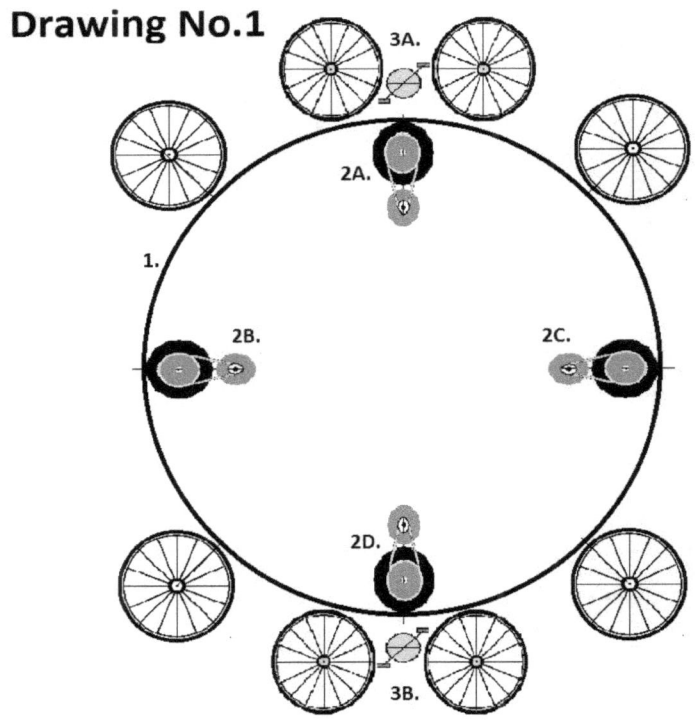

Drawing No.2: *Power Wheel without alternator*
Shows a side view of the Power Wheel without the alternator.
  A.) 1' dia. Pneumatic tire
  B.) Axle and attaching cap
  C.) 1' dia. Pulley
  D.) Adjustable alternator harness
  E.) 2 guide lines with brackets
  F.) 3.5' fence pole
  G.) 3.0' support post

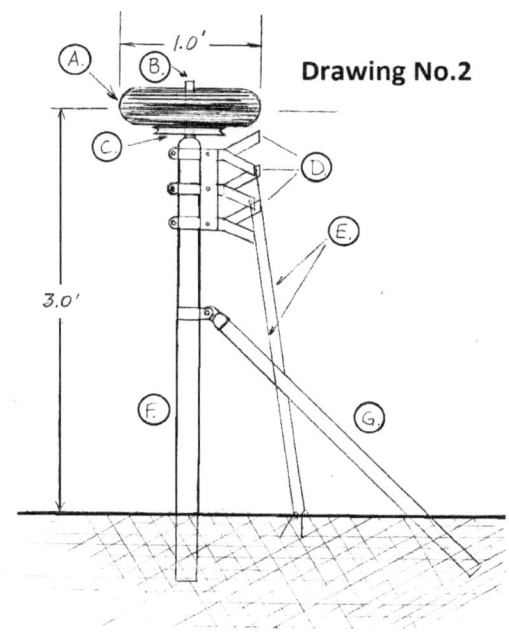

Drawing No.3: *Wheelbend Quadtrain*
Figure 1 show a top view of the Wheelbend section.
Figure 2 shows a front view of the Wheelbend section.
Figure 3 shows a side view of the Wheelbend section.
Each Wheel Quadtrain consists of the following:
A.) Fabricated 10x1' sheet metal cylinder section
B.) 10' long, 2" diameter structural support steel tubing
C.) 26" bike wheel
D.) 26" front fork

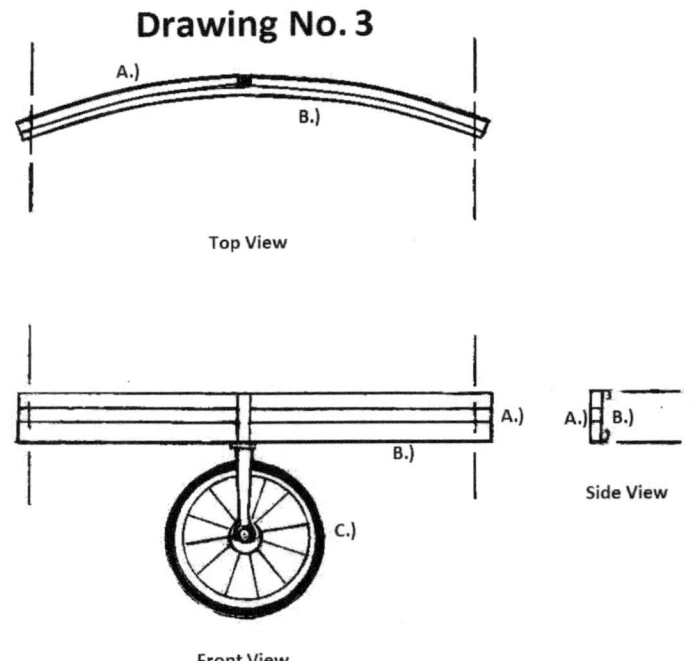

Drawing No.4: *Hoopbend Quadtrain*
Figure 1 show a top view of the Hoopbend section.
Figure 2 shows a front view of the Hoopbend section.
Figure 3 shows a side view of the Hoopbend section
Each Hoopbend Quadtrain consists of the following:
   A.) Fabricated 10x1' sheet metal cylinder section
   B.) 10' long, 2" diameter structural support tubing
   C.) 6" dia. Metal Hoop Ring

Drawing No.5: *Bike to Transfer Cylinder Connectors*

Drawing No. 4

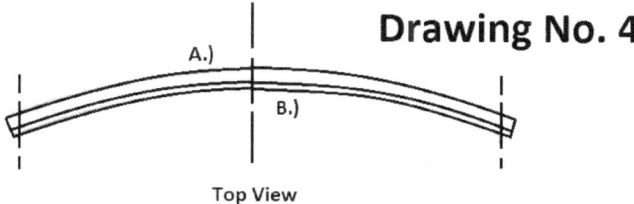

Top View

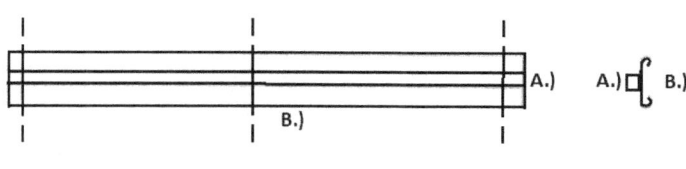

Front View

and Lock Bracket

Figure 1 show a side view of the Bike to Transfer Cylinder connector

Figure 2 shows a side view of the steel Hoop

Figure 3 shows a side view of the Bike Seat connector

Figure 4 shows a side and front view of the Lock Bracket

A.) Large Clasp
B.) Small Clasp
C.) Rope
D.) Shows a side view of the steel Hoop
E.) Shows a side view of the Bike Seat connector
F.) Shows views of the 2x2x6" U-Lock Steel Bracket

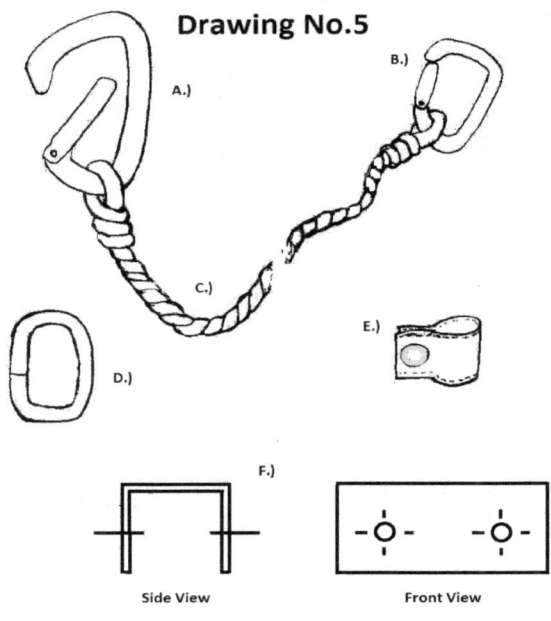

Drawing No.5

Detailed Description

(This basic configuration is the smallest version of the **Cyclonic Cylinder Speed Transfer System** design.)

Drawing No.1 shows a diagram of the Cyclonic Cylinder STS (Speed Transfer System) in its basic configuration. This basic configuration requires a minimum number of four PowerWheels and two bicyclists to operate efficiently.

The basic configuration of Transfer Cylinder consists of four Wheelbend sections (**Drawing No. 3**) and four Hoopbend sections (**Drawing No. 4**). Each of the eight sections measure ten feet in length. When fully assembled, the Transfer Cylinder measures eighty feet in circumference.

The device leverages the momentum of speeding bicyclist by attaching the bicycle by means of a specified length of rope or strapping to a rotating sheet metal cylinder. This cylinder rotates on a minimum of four bicycle wheels. The large diameter cylinder is held in place by a minimum of four power wheels.

Drawing No.2 shows the power wheels with reinforced posts serves three specific functions. First they are placed at equal distance from center to form a circular perimeter of specified circumference to sufficiently stabilize the integrity of a rapidly rotating sheet metal cylinder. Secondly, it must secure one alternator to itself for belt and pulley harnessing. Thirdly, the power wheel operates with the transfer cylinder to transfer the kinetic energy of the bicyclist's angular momentum and speed to the angular velocity of the alternator's rotor.

The elimination of spokes for stabilization of the

rotating Transfer Cylinder reduces the area required for utilization. This design feature creates usable space that may be occupied by certain structures or activities. For example, the space could be occupied by a small building or storage facility. Other suitable structures could be the geodesic dome or teepee. Also a tower for electrical distribution could also be fitted within the circular area of the unused space.

As the bicyclists ride at the optimal speed, the rotating Transfer Cylinder transfers the speed of tandem bicyclists to the PowerWheels, which in turn spin alternators by a belt and pulley to generate electricity. Most importantly, the **Cyclonic Cylinder STS (Speed Transfer System)** uses power of the speeding bicyclist's leveraged momentum to maintain the optimal speed for the longest possible time. The leverage is realized by the motion of opposing bicyclists traveling in a clockwise or counter-clockwise direction to generate the torque necessary to rotate the Transfer-Cylinder

Dimensions of the
Cyclonic Cylinder STS (Speed Transfer System):
Circumference: 80 feet
Radius: 12.73 feet
Height: approx. 3.5 feet

Assembly Instructions.

Step #1.) Determine suitable location to establish the point of center.

For the area to be suitable for the safe operation of the **Cyclonic Cylinder STS (Speed Transfer System)**, it must be free of obstacles like rocks and shrubs. The suitable area must also be relatively level and flat plane. Also for aesthetic consideration, the ideal area would be situated as near to scenic surroundings as practical.

The transfer cylinder requires a 25' diameter and the bicyclists require additional footage or about four additional feet. That means, you'll need at least 29'-30' square foot area.

Step #2.) Once a location has been determined, find the center or pivot point.

From this point, measure out to locate the positions for each PowerWheel posts.

In addition to locating the equal point from center for each PowerWheel. The PowerWheels must also be distanced equally from each other.

Step #3.) Set the PowerWheel support post in designated location.

Each PowerWheel support post should be secured using the same methods and techniques to properly set steel fence posts. All play or wobble in the post setting should be minimized if not eliminated. If needed, additional guide wire can be used to tension up the setting.

Step #4.) Assemble PowerWheel.
Step #4A.) Attach alternator harness to post.

(See Drawing No.2: Power Wheel without alternator Figures A, B & C for the following.)

Step #4B.) Attach axle to post.

Step #4C.) Secure one foot dia. Rubber Tire with attached pulley.

(See Drawings 3 and 4 for the following.)

Step #4D.) Assemble Transfer-Cylinder. Place all eight fully assembled sections around the perimeter formed by the power wheel posts. Secure the sections with the metal brackets.

Step #4E.) Assemble alternators and connect to junction box. Secure one alternator to each post with harness. Attach wiring from alternator to junction box.

Step #4F.) Attach bicycles. Secure strap around seat post. Using the (x) foot length rope to tie the bike and then to tie the transfer cylinder rings.

Step #4G.) Begin cycling power. Monitor power for consumption and transmission. The best mode for the smallest version, which is 80' circumference, is between two and four bicyclists, riding in tandem, and operating eight power wheel posts. The bicyclist should maintain a speed between 15 mph to 30 mph.

This is only the first model, and as such subject to profound improvements and refinements. But this is more than just an engineering innovation, because if it proves as effective as theorized it could form the nucleus for a social movement that will free the indentured multitude from the rapacious greed of the energy hegemony.

Chapter Four

The New Ordered World is Now.

It's becoming increasing difficult to disguise the fact that a majority of this planet's human population now finds itself in an unenviable state of collective servitude. For everyday, this vast multitude is expected to pay out costly sums in hidden fees and unfair taxes. Even worse, this mass state of exploitive indenture-ship, enabled and maintained by the insatiable greed of energy profiteers festers and grows more onerous ever since it's dubious inception so long ago during the age of steam!

Even Nikola Tesla, the electrical wizard who single-mindedly illuminated the 1893 Chicago Exhibition with his patents on alternating current motors and alternators had railed against this greedy construct. And he would remain determined and steadfast in his personal quest toward discovering a free source of electrical power. An abundant source of power that would end the monopolistic reign of utility companies and their unabated exploitation of the uninformed masses to the end of his life.

But why take my word for it! Read for yourself the following passages from an article that appeared in the **March-1896** edition of the World Sunday Magazine that was aptly titled, "Earth Electricity To Kill Monopoly." Some would say this harkens of a clarion call demanding some form of utility debt relief for the working masses.

For the opening paragraph unabashedly declares, "The end has come to telegraph and telephone monopolies with a crash...all the other monopolies that depend on power of any kind will come to a sudden stop."

It's almost comical at how absolutely dead wrong the author was proven to be, especially when considering our current deplorable servile condition. Which is now well over a century later.

But this was just a tame prelude to the very next paragraph that stated unequivocally, "It means that if Nikola Tesla succeeds in harnessing the electrical earth currents and putting them to work for man there will (be) an end to oppressive, extortionate monopolies in steam, telephone, telegraphs..and the grasping millionaires who have for two decades milked the people's purse with electrical fingers..."

A few paragraphs down, the article ends with a few more revealing tidbits, "Electricity will be as free as the air. For the privilege of its use legislatures will not have to be bribed or men corrupted at the polls, and public boards will not have to be seen to bestow exclusive franchises upon corporations organized to use public property for purposes of private gain, and make the people pay the original cost of their investment and excessive charges for service...Monopolies for purveying steam power too will be forced to capitulate to free electricity...The successful adaptation of Tesla's discovery will administer a death-blow to the most galling slavery that has yoked the activities of men to the treadmill of monopoly."

The Ring of Power draws current directly from the

Nikola Tesla's gleaming alternating current aspirations to produce plentiful power for all humanity.

But more specifically, the Ring of Power is a "leveraged momentum machine" and combines most of the known forces in the Universe to deliver electrical power to those who can least afford it.

Let me reiterate what was stated for unmistakable clarification. That is to say, human beings have been endowed with the capacity to be energy self-sufficient. It is only through the artificial constraints of a greedy society that we find ourselves forced to owe our existence to these global, energy-hoarding profiteers. We have been cajoled and berated into accepting the false premise that the ability of man to create their own power is a futile attempt at best and should therefore be avoided at all costs.

These money-minded disciples of greed are quick to throw out comparative energy costs and nascent technology limitations. They will throw calculations in your face to demonstrate just how relatively little a human can generate with current technologies. But this is all just what it is, words from the master declaring to his slaves that everything is fine just as they are. And for the sake of the burgeoning global economy, we should keep our ignorant, cotton-picking hands out of their highly lucrative free (to exploit) markets.

And that's why it's incumbent upon those not fully-invested with this lopsided socioeconomic construct to make the necessary efforts to rediscover this great potential of the human body. And it's not simply about leveraging muscle power, it's about leveraging momentum. You know, the reason why collisions in full contact sports like football and hockey can be so devastating.

And it's with this thought in mind that I propose the following calculation: People-Power equals Clean Energy plus Clean Environment multiplied by a healthier and wealthier population as the sum numerator. Then divide this sum by a shrinking unemployment and national deficit.

What you would ultimately end up with should finally equate to a truly financially and physically sound global population. Of course, the goal is for the denominator to shrink to zero and then cancel itself out.

But in all seriousness, how can we make this calculation a reality and not just bold words inked on paper? To answer that I would offer the following observation. We are fast approaching the fiftieth anniversary of man's first visit to moon. But fifty years before that historic Apollo Moon Launch, the mere thought of going to the moon was considered fanciful science fiction. Now, it's just a matter of time before we visit our crimson celestial neighbor, the planet Mars.

But brain power does not preclude our daily requirements of physical activity. Its not by chance that our minds are enhanced and leverage by our capacity to manifest willpower through our primary force actuators. That is, our legs, arms, hands and feet.

Yet, once again, by circumstance of man's own contrived mechanical and electrical devices, social norms, taboos and other egregious forms of societal malfeasance the masses have been made to believe that we must pay to stay fit and healthy.

This cost to society is now accepted as a way of life. But what if we could turn the table on this greedy calculation and pay people for staying physically fit and healthy. Would they not be more productive and happier? And wouldn't the money they earned ultimately circulate back into economy?

But where would this money come from to pay people to stay fit, trim and healthy? Well, basically, the money comes from the power they generate. Which is why I truly believe that it's not an overstatement to claim that the New World Power movement has the potential to genuinely free the masses from the shackles of costly energy and establish a New Ordered World (NOW).

To do this, NWP will offer powerful incentives to motivate it's participants to become fit and stay in shape! This long term lifestyle change will be accomplished through generous monetary and recognition rewards for their exercise efforts. But this is only the start of what New World Power hopes to achieve for the betterment of our global society and not just a privileged few.

For one of the more ambitious goals of NWP will be to offer the able-bodied unemployed and or the homeless the opportunity to earn gainful wages through wholesome, body-building exercise.

But again, the movement will not stop there because NWP, in conjunction with its "Fight Fat & Live Fit," program, will offer healthy and wholesome meals at little or no cost to all NWP members.

Also in addition to this, NWP plans to provide onsite access to personal trainers, physical therapists and medical experts who can provide a full spectrum of healthy living expertise. And while it may not fall within the scope of NWP, we will work with local, state and federal agencies to see that these people will have full access to other housing, employment and educational opportunities.

Contained in the New World Power movement exists the momentous spark for the societal transformations capable of evolving our society beyond ways presently imaginable. This growing confidence stems from the knowledge that people should have the means to be energy self-supportive and truly free from the tyrannical greed of global corporate governance.

To achieve NOW, the New World Power movement, like any worthwhile effort that truly benefits mankind will be an epic struggle. But it is a good fight worth fighting because while however lofty may be NWP's goals and objectives, once reached, these transformations will unleash a renewed wave of hope in humanity as a noble collective soul imbued of divine aspirations.

So how does NWP plan to wean our society from it's destructive addiction to expensive foreign energy sources? First off, NWP will not rely on any one product, technoloy nor strategy because it will take everything necessary for the effective realization of NOW.

However at the time of this writing, the lead objective calls for obtaining adequate funding to construct a working prototype of the next generation, "Ring of Power." And also to conduct a series of tests and trials for determining the currently unknown, though theorized, operating parameters of this new device for generating electricity.

If everything moves forward accordingly, we should expect to see the first trials for the working prototype begin no later than July 16, 2015.

And by the 50th anniversary of Apollo Moon landing,

the "Ring of Power," will become a stabilizing force in the global affairs of our planet Earth.

Bibliography

Landels, J.G., **Engineering in the Ancient World**, Berkeley & Los Angeles, University of California Press, 1978.

The Random House Encyclopedia, New York, Random House, Inc., 1977

Ronan, Colin A., **Science: Its History And Development Among The World's Cultures**, New York, Facts on File, 1982.

Tilley, Donald E., **Contemporary College Physics**, Menlo Park, California, The Benjamin/Cummings Publishing Company Inc., 1979.

www.ingramcontent.com/pod-product-compliance
Lightning Source LLC
Chambersburg PA
CBHW070721180526
45167CB00004B/1573